Everyday Mathematics®

The University of Chicago School Mathematics Project

Mathematics at Home
Book 4

Wright Group

The McGraw·Hill Companies

Wright Group

Send all inquiries to:
Wright Group/McGraw-Hill
P.O. Box 812960
Chicago, IL 60681

ISBN 0-07-604522-6

2 3 4 PBM 09 08 07 06

The **McGraw·Hill** Companies

Authors

Jean Bell
Dorothy Freedman
Nancy Hanvey
Deborah Arron Leslie*
Ellen Ryan
Barbara Smart*

* Third Edition only

Consultant

Max Bell
University of Chicago School Mathematics Project

Third Edition Early Childhood Team Leaders

David W. Beer, Deborah Arron Leslie

Editorial Assistant

Patrick Carroll

Contributors

Margaret Krulee, Ann E. Audrain, David W. Beer

Contents

Introduction

This is the last *Mathematics at Home* book for *Kindergarten Everyday Mathematics*. We hope you are still looking back and repeating favorite activities from earlier books, extending the activities, adapting them by changing the numbers, and adding your own ideas.

We also hope you will keep the *Mathematics at Home* books in mind this summer as you share time with family and friends and find ways to take advantage of the mathematics opportunities that are all around you. This book has some specific suggestions for summer activities. Our basic idea still holds: Let mutual interest and enjoyment guide you in exploring mathematics with your child.

Mathematics Outdoors

Measures, Counts, and Patterns

Measure an outdoor area (a yard, playground, or baseball diamond, for example) with your feet or with paces (natural walking steps). Later, you might use measuring tools such as a tape measure, yardstick, or length of string.

Measure how far you can throw a ball or do a long jump.

Count the rungs on the monkey bars or the ladder steps on a slide.

Use numbers and count as you play hopscotch and sing jump-rope counting rhymes.

Make patterns out of natural objects such as small stones, leaves, or sticks. Ask someone to extend your pattern.

Play *Follow the Leader* with patterned movements like this: step, step, hop; step, step, hop; and so on.

Playground Math

Time how long it takes to do an outdoor activity such as running down the block or crossing the monkey bars. (Or, ask someone else to time the activity while you do it.)

Create an obstacle course on the playground. Use position words to give directions and use a timer to see how long it takes to complete it. You might want to measure parts of the course with your feet or other body parts or with standard measuring tools.

Use comparison words like *higher* and *tallest* to describe how high the swing can go or how tall the jungle gym is.

Give an empty swing a push. How far can you count before it becomes perfectly still?

Shapes and Symmetry

Look for shapes in the park and on the way home. How many different shapes did you find?

Look closely at the bodies of butterflies and other insects you find. Are they symmetric (look the same on both sides)? Do you see patterns? Describe what you notice.

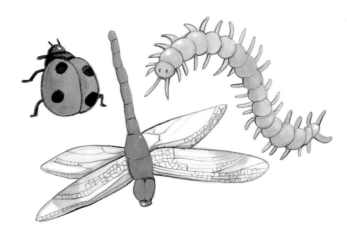

Shadow and Moon Explorations

On a bright sunny day, have someone trace your shadow on the sidewalk with chalk. Try this in the morning, at noon, and late in the afternoon. Measure the shadow with your feet or another measuring tool. How does the size change at different times of the day? Are there any other changes?

Watch the moon for many nights in a row. Notice the different shapes the moon appears to make. Do you notice a pattern in the way the moon changes? How many days do you think it will take until you see a full moon (a full circle) again? You might want to sketch on a calendar how the moon looks each night.

Counting

Keep Going

Don't let the number 100 be a barrier. Count past it! Try counting from different starting numbers such as 81, 92, or 68.

Continue to skip count by 2s, 5s, and 10s. Add a rhythm or familiar tune and turn it into a counting song!

How Many in a Handful?

Put some small objects (pennies, blocks, cereal) in a bowl. Grab a handful. Estimate how many there are, then count. Were you right? Have someone else grab a handful. Do they have the same amount? Do they have more or less? How many more or less? Try again, this time putting the objects in piles of 2, 5, or 10 and using skip counting to count them.

7

Name Collections

Ways to Show Five

Help your child count out 5 pennies. Together, arrange the pennies in several different ways: a straight line, close together, far apart, by 2s with one left over, in a group of 3 and a group of 2, and so on. Discuss with your child that however you arrange them, the number is still 5. Try doing this with other numbers, too! Your child might like to record some of the arrangements with pictures, numbers, or words.

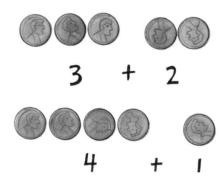

$$3 \quad + \quad 2$$

$$4 \quad + \quad 1$$

Groups of 5

Name Collection Cookies

Use a roll of refrigerated sugar-cookie dough and small candies or baking chips in two or more colors to make cookies that show different number combinations. For example, you might decorate a cookie with 3 blue candies and 1 red candy, or with 2 chocolate chips and 2 butterscotch chips. Sort the cookies by the total number of decorations on each. Read your cookie number before you eat! For example, "My cookie shows 3 + 1 = 4."

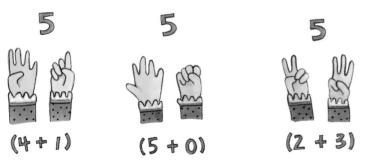

Finger Math

Use your fingers to show different ways you can make 4, 5, 6, . . . Finger math is fun!

Patterns

Pattern Projects

Look for patterns on clothes. Copy a pattern you find onto a sheet of paper using crayons or markers.

Pick two or more of the shapes shown here. Draw one or more patterns using these shapes. Remember, you can turn the shapes upside down and sideways too!

10

Measuring and Comparing

Mix and Measure

Make some modeling dough with your child. Talk about the different sizes of the measuring cups and spoons as you work together. Compare the amounts as you mix up a batch. (The modeling dough made from the recipe on this page and the dough from the recipe on the next page are nontoxic but are not intended to be eaten.)

No Cook Modeling Dough

1 cup salt

3 cups flour

2 tablespoons cooking oil

About $\frac{3}{4}$ cup water (enough to make a semi-firm ball)

Food coloring or unsweetened, flavored drink mix

Mix the ingredients. Store the modeling dough in the refrigerator.

Cooked Modeling Dough

$2\frac{1}{2}$ cups flour

$\frac{1}{2}$ cup salt

2 packages unsweetened, flavored drink mix

3 tablespoons cooking oil

2 cups boiling water

Mix the dry ingredients. Add the oil and boiling water and stir until a soft ball forms. (Add extra flour if necessary.)

Store the modeling dough in the refrigerator.

Modeling Dough Math

Now have fun "doing math" with your modeling dough!

◆ Form numbers from long "snakes" of dough.

- Make spheres of different sizes. Line them up from smallest to largest. Then reverse them and line them up from largest to smallest.
- Make two spheres that are the same size, then roll one into a snake. How long is the snake? Do you think the sphere and the snake weigh the same amount?

Size It Up

Use math language to compare sizes when you go shopping for new clothing or new shoes. Encourage your child to describe a shirt or shoes using language such as *too big, too small, not big enough, too short, too long,* and so on.

Use a string or piece of yarn to measure how long your child's arm is. Attach a label with his or her name. Then help your child measure other family members' arms—or maybe the cat's tail or dog's leg. Compare the strings and discuss which ones are the *same, longer,* or *shorter.*

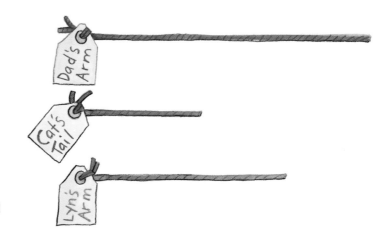

Equal Sharing

Family Note

Children's earliest exposure to mathematical division and fractions is usually through equal sharing (dividing any whole object or group of objects into 2 or more equal parts or groups). Children usually understand that dividing something equally is the fairest way to share a treat.

Fair Shares

Take a small group of objects, such as pennies, beans, or popcorn, and divide them into 2 equal groups, 3 equal groups, and so on. An interesting question to discuss is "What can be done with the leftover objects?"

Equal Sharing in Children's Books

Read *The Doorbell Rang* by Pat Hutchins and discuss the problem of sharing the food fairly. You might enact this story, or a variation of it, at home! See page 22 for suggestions of other books that deal with fractions and equal sharing.

Equal groups

Leftovers

Time

Fun and Work Time

Guess how much of a room you can clean up in 2 minutes. Set a timer and try cleaning for 2 minutes. Did you do more or less than you thought you would?

Let your child be responsible for turning the television set on at the correct time to watch a favorite TV show. Explain where the clock hands will be when the show begins. Drawing a picture or setting the hands of a play clock for the designated time can be helpful.

Ideas for Summer Activities

Summer Calendar

Set up a calendar to keep track of summer events. You might use it for any or all of the following activities:

◆ Figure out how many months, weeks, and days will pass before school starts again. Set up a way to count up or back to the first day of school.

◆ Mark special events, such as birthday parties or family outings or trips.

◆ Mark when your library books are due.

◆ Record what the moon looks like each night. (See page 6 in this book.)

Plant some watermelon seeds and watch them grow. You can keep track of their growth by measuring them. You might want to plant corn, beans, and radishes because they grow quickly.

Watermelon Math

When you get watermelon at the grocery store, try to guess how much it weighs. Check your guess at the produce scale (or by reading the label). As an adult cuts it up, estimate how many seeds are in different slices. Count and compare estimates as the slices are eaten.

Sand Activities

"Do math" at the beach. Make footprint or handprint patterns in the sand. Describe the patterns! Take turns copying or extending patterns with a friend.

Take a variety of sizes of nonbreakable buckets or cups to the beach to pour sand or water into. Count how many cups it takes to fill a larger bucket. Tally them and then write the number in the sand.

Sink or Float?

Explore whether these objects (and others) sink or float: eye dropper, funnel, plastic measuring cup, baster, sponge, and a bottle of bubble bath soap. Test them in a wading pool or in the bathtub. How many items float? How many sink? Think of a way to record and display the information.

Continue to experiment with these and other materials. Do the same things always float or sink? Can you make sinking things float or floating things sink? For example, try empty bottles versus full ones; dry and wet sponges; and a sheet of aluminum foil compared to foil wadded up in a ball or shaped into a boat.

Lemonade Stand Fun

A lemonade stand can provide hours of enjoyment and a variety of mathematical experiences: measuring, estimating, counting, and handling money.

Decide how much lemonade to make. About how many cups do you want to try to sell? (Smaller-size cups may work best.)

Here is a recipe for lemonade using fresh lemons. The amount of water and sugar can be varied to suit your taste.

Fresh Lemonade

$\frac{1}{2}$ cup fresh lemon juice (about 2 or 3 lemons)

$\frac{1}{3}$ cup sugar (or to taste)

About 4 cups of cold water

Mix ingredients well until sugar dissolves completely.

Makes about 5 servings.

Think about how much to charge for each cup so you can pay for the materials and make a little extra money.

At the end of the sale, make piles of each type of coin or bill you earned and count your money together with a family member. You might want to use a calculator.

Zero-trash Picnic

Think about how you can help keep your neighborhood clean. Plan a zero-trash picnic and consider the following:

- ◆ What should we eat?
- ◆ How much food will we need?
- ◆ How can we pack our food so we have nothing to throw away when we are done? What containers might we use?

Some Books for Children

Big Numbers

Gag, Wanda, *Millions of Cats* (Putnam Juvenile, 2004)

McCourt, Lisa, *100th Day of Bug School* (HarperFestival, 2005)

Pallotta, Jerry, *Count to a Million* (Scholastic, 2003)

Pinczes, Elinor, *100 Hungry Ants* (Houghton Mifflin, 1999)

Ross, Tony, *Centipede's One Hundred Shoes* (Henry Holt & Company, 2003)

Schwartz, David M., *How Much Is a Million?* (HarperTrophy, 1993)

Clocks and Time

Carle, Eric, *The Grouchy Ladybug* (HarperCollins, 1996)

Cousins, Lucy, *What's the Time, Maisy?* (Walker Books, 2002)

Hawkins, Colin, *What's the Time, Mr. Wolf?* (Egmont Books, 2004)

Herbst, Judith, *Big Hand, Little Hand: Learn to Tell Time!* (Barron's Educational Series, 1997)

Hutchins, Pat, *Clocks and More Clocks* (Aladdin, 1994)

McCaughrean, Geraldine, *My Grandmother's Clock* (Clarion, 2002)

Older, Jules, *Telling Time* (Charlesbridge Publishing, 2000)

Fractions

Hutchins, Pat, *The Doorbell Rang* (HarperTrophy, 1994)

Leedy, Loreen, *Fraction Action* (Holiday House, 1996)

McMillan, Bruce, *Eating Fractions* (Scholastic, 1991)

Wood, Don and Audrey, *The Little Mouse, The Red Ripe Strawberry, and the Big Hungry Bear* (Child's Play International, 1984)

Measures and Size

Jenkins, Steve, *Actual Size* (Houghton Mifflin, 2004)

Leedy, Loreen, *Measuring Penny* (Henry Holt and Co., 2000)

Pinczes, Elinor J., *Inchworm and a Half* (Houghton Mifflin, 2003)

Money

Brisson, Pat, *Benny's Pennies* (Dragonfly Books, 1995)

Gill, Shelley & Deborah Tabola, *The Big Buck Adventure* (Charlesbridge Publishing, 2000)

Silverstein, Shel, "Smart" from *Where the Sidewalk Ends* (HarperCollins, 2004)

Viorst, Judith, *Alexander, Who Used to Be Rich Last Sunday* (Atheneum, 1978)

Williams, Vera, *A Chair For My Mother* (HarperTrophy, 1994)